Weathering and Erosion
and the Rock Cycle

Joanne Mattern

The Rosen Publishing Group's
PowerKids Press™
New York

Published in 2006 by The Rosen Publishing Group, Inc.
29 East 21st Street, New York, NY 10010

First Edition

Editor: Melissa Acevedo
Book Design: Ginny Chu

Illustrations: Pp. 6 and 7 by Ginny Chu.
Photo Credits: Cover and title page © Hubert Stadler/Corbis; p. 4 (top) © Darrell Gulin/Corbis; p. 4 (bottom) Richard Hamilton Smith/Corbis; p. 8 (top) © Michael DeYoung/Corbis; p. 8 (bottom) © W. Cody/Corbis; p. 10 (left) © Mark A. Johnson/Corbis; p. 10 (right) © Randy Faris/Corbis; p. 12 (top) © Royalty-Free/Corbis; pp. 12 (bottom), 18 © David Muench/Corbis; pp. 14, 15 © Steve Kaufman/Corbis; p. 16 © Danny Lehman/Corbis; p. 17 © Wolfgang Kaehler/Corbis; p. 19 (left) © Wayne Lawler; Ecoscene/Corbis; p. 19 (right) © Craig Aurness/Corbis; p. 20 © Najlah Feanny/Corbis Saba; p. 21 © John Henley/Corbis.

Library of Congress Cataloging-in-Publication Data

Mattern, Joanne, 1963–
    Weathering and erosion and the rock cycle / Joanne Mattern.— 1st ed.
        p. cm. — (The shaping and reshaping of earth's surface)
    Includes index.
    ISBN 1-4042-3198-6 (lib. bdg.)
    1. Weathering—Juvenile literature. 2. Erosion—Juvenile literature. 3. Geochemical cycles—Juvenile literature. I. Title.

    QE570.M369 2006
    551.3'02—dc22
                                    2005002366

Manufactured in the United States of America

Contents

When plants become trapped inside a crack in a rock, like in the picture at left, it weakens the rock. As the plant grows, it further breaks down and cracks the rock. Plants growing inside a rock is one of the major causes of weathering.

Sometimes weathering causes layers of rocks to peel off like an onion skin. The mountain of Corcovado, over the harbor of Rio de Janeiro, Brazil, is an example of this weathering.

Right:
When ice gets inside a crack in a rock, it can break pieces off the rock and weather it.

CONTENT SKILL: Weathering and Erosion

Weathering and Erosion

What Are Weathering and Erosion?

Rocks are hard and strong, but they do not stay that way forever. Forces like wind and water break down rocks through the processes of weathering and erosion.

Weathering is the process that breaks down rocks. Many things cause weathering, including climate changes. Erosion breaks rocks down further and then moves them. Forces like wind and water move the rock pieces. They mix with matter like sand to become sediment. Weathering and erosion help shape Earth's surface. They are part of a process called the rock cycle.

Weathering is the process that breaks down rocks. Erosion breaks rocks down further and then moves them.

Weathering and Erosion in the Rock Cycle

The rock cycle is the process through which rocks are broken down to create new ones. Weathering and erosion are important parts of the rock cycle. This cycle has been shaping and reshaping Earth for millions of years.

The rock cycle begins when hot, liquid magma rises to the surface of Earth. Once there, it cools and hardens into igneous rocks. Over time weathering wears down the igneous rocks. As these rocks erode, they get mixed with other matter to create sediment. The sediment then moves into various bodies of water through the process of erosion, where it settles into layers. In time these layers

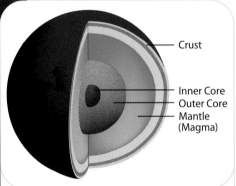

Crust

Inner Core
Outer Core
Mantle
(Magma)

Earth is made up of many layers. The top layer, called the crust, is made of rock. Under the crust is a layer called the mantle. It is made of a hot liquid called magma. Below this layer is Earth's core. The outer core is made up of melted metals. The inner core is a ball of solid metal.

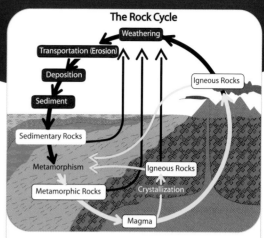

The Rock Cycle

This diagram shows the process called the rock cycle, which creates and destroys rocks. Weathering and erosion help form sedimentary rocks by creating and moving sediment.

form sedimentary rocks. Metamorphic rock forms when a great deal of Earth's heat or pressure comes in contact with the minerals in igneous and sedimentary rocks and changes them. The rock cycle continues when magma melts rocks that are pushed below the surface of Earth through earthquakes and other movements. Weathering and erosion play a major part in the rock cycle by breaking down rocks so new rocks can form.

Weathering and erosion are important parts of the rock cycle. This cycle has been shaping and reshaping Earth for millions of years.

The climate in this cave is moist and hot because of the heat of the Sun. The Sun chemically weathers rocks by changing their minerals.

As rain soaks into the ground, it picks up chemicals, such as nitrogen, from the soil around it. These chemicals can weather the rocks on Earth's surface by adding or taking away minerals from the rock's chemical makeup.

Right:
Rust causes chemical weathering. This rock is turning red because rust has built up on it.

What Causes Weathering?

Chemical Weathering

There are two kinds of weathering. Chemical weathering occurs when elements like oxygen cause substances in the rock to go through changes that affect its chemical makeup. These changes break the rock. For example, some rocks contain the mineral iron. When oxygen in the air touches iron, rust forms. Rust weakens the rock.

Climate can also cause chemical weathering. Rocks weather more than three times faster in hot, wet climates than in cool dry climates. This is because the high temperatures and large amount of water cause the rock's minerals to change.

Chemical weathering occurs when elements like oxygen cause substances in the rock to go through changes that affect its chemical makeup.

Physical Weathering

Physical weathering is the second kind of weathering. Forces like wind, water, and ice weaken the surface of a rock and then break it down in this type of weathering.

The wind in this picture is blowing snow over some rocks. The snow will weather the rock by breaking off bits and creating cracks.

A gentle breeze might not feel powerful to you. However, when it blows against a rock over a long period of time, it can cause physical weathering. Wind moving over a rock scrapes off tiny bits. It can also blast rocks with small pieces of matter like sand that weaken and break down the rock's surface.

Water is a force in physical weathering. The force with which ocean waves hit rocks weakens the surface of the rocks, causing pieces to break off.

Water can also physically

weather rocks. Physical weathering through water works in many ways. When water flows over rocks, it can remove pieces of the rock. The force of waves crashing against rocks at the beach can cause the rock to break apart.

Wind is a powerful instrument in the process of physical weathering. Wind blowing in Japan can carry small bits of matter more than 1,000 miles (1,609 km) from China.

Water can also get trapped inside a rock. When the trapped water pushes against the rock to get out, it causes cracks. Sometimes the water that is trapped inside a rock freezes and becomes ice. The ice spreads and pushes against the rock, breaking it apart.

A gentle breeze might not feel powerful to you. However, when it blows against a rock over a long period of time, it can cause physical weathering.

Water is a major cause of erosion. When water hits the surface of a rock, as shown here, it breaks off weak pieces and carries them away.

As water flows downhill, carrying away bits of rock and other sediment, it can erode a small path into the soil. These paths are called rills. If enough water flows through the rill, it will grow into a stream.

Right:
In erosion, flowing water is transportation for rock bits, sand, and other matter.

What Causes Erosion?

Water

After weathering has broken down the rocks, rock bits mix with mud and other matter through erosion. A force like water carries them as sediment to the ocean. Water is key in erosion's transportation process.

The transporting process begins when rain falls. When the rain washes over the rock, it not only weathers the rock, but also carries away little bits of it. The rain carrying the rock pieces flows into streams and rivers. In time some of these rivers flow into the ocean, carrying the matter with them that settles into layers to create sediment.

Water is key in erosion's transportation process. The transporting process begins when rain falls.

13

Wind

Wind is another powerful force in the process of erosion. It plays a major part in shaping and reshaping Earth's surface. Wind is important because it can change the shape of rocks as it erodes their surface. There are two kinds of wind erosion. The first is called deflation. Deflation happens when the wind picks up sand, mud, and other matter. The heavier substances, such as rocks, stay behind as the wind carries away the smaller, lighter pieces of matter. This is how rocky deserts are created.

Rocky deserts are created through the process of deflation, which is a type of erosion. The Negev Desert, shown above, is located in Israel. Erosion created this desert's rocky landscape.

Abrasion is the second kind of wind erosion. Abrasion occurs when the wind carries away bits of sediment. The wind then blows those bits against a rock's surface. Abrasion sometimes forms tiny holes in a rock. It may also scratch a rock, forming tiny lines. Other times abrasion can polish the rock's surface until it is smooth and shiny. Sometimes, through the process of abrasion, the wind carves huge holes into soil and rocks. These holes are called blowouts. They can be hundreds of miles (km) long!

This close-up view of a piece of sandstone shows the spots where the rock has been eroded. The holes in the center, called blowouts, were created by strong winds over time.

Wind is another powerful force in the process of erosion. It plays a major part in shaping and reshaping Earth's surface.

Ice

A third powerful force of erosion is ice. Large pieces of ice called glaciers move over Earth. As they move they scrape Earth's surface and erode it, carrying away rock and sediment. Glaciers form in high mountains and near the snowy North Pole and South Pole. These places are so cold that not all the snow melts. Instead it piles up until it is so deep and heavy that it turns into ice over time. This is how a glacier is formed.

This glacier, located in Alaska, is already eroding through rock.

Glaciers can be up to several miles (km) thick!

Glaciers move very slowly over Earth. A glacier may only move a few inches (cm) in a day. However, they are so hard and heavy that they erode everything in their path as they scrape over the ground. Over

Moraines are piles of sediment that are left behind by glaciers.

thousands of years, a glacier can wear away huge rocks. It can create a valley where a mountain used to be. It can also create a moraine with the sediment it leaves behind. For this reason glaciers play a major role in the process of erosion and the rock cycle.

Large pieces of ice called glaciers move over Earth. As they move they scrape Earth's surface and erode it, carrying away rock and sediment.

17

Changing Earth's Surface

New Landforms

Erosion, through the forces of water and wind, is constantly changing Earth's surface by wearing down rocks and moving sediment around. These forces shape rocks and create new landforms.

Fast-moving water shapes Earth by eroding rock to create canyons and valleys. When a river flows into the ocean, it can leave sediment behind and create a plain called a delta.

The Grand Canyon was created through erosion. More than five million years ago, the Colorado River cut through the rock to create this huge hole in the ground.

Through the process of erosion, wind can also shape Earth's surface. Wind creates odd-looking rock formations. These formations occur

This delta is located in the Bismarck Sea in New Guinea.

This valley, located in Kanton Bern, Switzerland, was created by a glacier.

because the wind cannot carry heavy sand more than 6 feet (2 m) above the ground. The sand carried by the wind erodes the bottom part of the rock. This creates rocks that look like mushrooms!

Glaciers also shape Earth through erosion. As the climate gets warmer, glaciers will move and leave piles of sediment behind. The sediment piles up into hills called moraines. Glaciers also carve valleys as they creep across Earth's surface and scrape the rock away.

Erosion, through the forces of water and wind, is constantly changing Earth's surface by wearing down rocks and moving sediment around.

Bad Effects

Erosion is an important part of the rock cycle and has helped shape Earth's surface. However, erosion can also cause a great deal of harm. When water moves in large amounts and at a quick speed, it can cause a flash flood. These floods can erode or wash away everything in their path in a very short time. They can also harm Earth's surface by destroying soil. Floods can wreck buildings and can kill people.

Too much water can also cause sinkholes. A sinkhole forms when water erodes the rock underground. In time

The picture above shows the kind of harm floodwaters can do. These houses in East Grand Forks, Minnesota, were completely destroyed because of large amounts of floodwater. The people in the town tried to hold the waters back with sandbags, but the flood was too powerful.

This sinkhole in Richmond, Virginia, was created by a hurricane. A hurricane is a storm with strong winds and heavy rain. The heavy rain formed this sinkhole.

there is not enough rock to support the surface, and the surface collapses.

Wind can also harm Earth's surface by blowing away soil. This is especially bad when wind blows away the soil that is used for farming. Many times people make this erosion worse by cutting down trees or letting cattle eat all the plants. Plant roots help hold the soil in place. Without the roots wind can easily blow the soil away.

When water moves in large amounts and at a quick speed, it can cause a flash flood.

The Importance of Weathering and Erosion

Weathering and erosion have shaped Earth for millions of years. Without these processes the rock cycle could not continue. Weathering and erosion take bits of rock and other matter to create sediment, which is needed to form sedimentary rocks. These processes also shape Earth's surface by creating deltas, canyons, and other formations.

Weathering and erosion can affect people as well as Earth. They can create new landscapes. However, they can also cause floods and soil erosion. These things can harm people and Earth.

Weathering and erosion are a major part of the rock cycle. Without these two processes, rocks could not break down to form new rocks, and Earth's surface would never change.

Glossary

abrasion (uh-BRAY-zhun) The rubbing together of rock and tiny bits of rock.

canyons (KAN-yunz) Deep, narrow valleys.

chemical (KEH-mih-kul) Having to do with matter that can be mixed with other matter to cause changes.

deflation (dee-FLAY-shun) The movement of soil, sand, and dust by wind.

delta (DEL-tuh) A pile of earth and sand that collects at the mouth of a river.

erosion (ih-ROH-zhun) The wearing away of land over time.

glaciers (GLAY-shurz) Large masses of ice that move down a mountain or along a valley.

igneous rocks (IG-nee-us ROKS) Hot, liquid, underground minerals that have cooled and hardened.

metamorphic rock (meh-tuh-MOR-fik ROK) Rock that has been changed by heat and heavy weight.

minerals (MIN-rulz) Natural elements that are not animals, plants, or other living things.

moraine (muh-RAYN) A hill of earth and stones left behind by a glacier.

physical (FIH-zih-kul) Having to do with natural objects.

sediment (SEH-dih-ment) Gravel, sand, or mud carried by wind or water.

sedimentary rocks (seh-dih-MEN-teh-ree ROKS) Layers of stones, sand, or mud that have been pressed together to form rock.

substances (SUB-stan-siz) Any matter that takes up space.

transportation (tranz-per-TAY-shun) A way of traveling from one place to another.

weathering (WEH-thur-ing) The breaking up of rock by water, wind, and chemical forces.

Index

Web Sites

Due to the changing nature of Internet links, PowerKids Press has developed an online list of Web sites related to the subject of this book. This site is updated regularly. Please use this link to access the list:

www.powerkidslinks.com/sres/weatheros/